# PETER LAYTON & FRIENDS

celebrating london glassblowing

# PETER LAYTON & FRIENDS

## celebrating london glassblowing

compiled by Peter Layton

HALSGROVE

First published in Great Britain in 2006

Text Copyright © 2006  Peter Layton and individual contributors
Image Copyright © 2006 resides with individual copyright holders

Unless otherwise stated all illustrated works are by Peter Layton

Image on previous page:  Mirage cased stone form, 2000, 25cm high

British Library Cataloguing-in-Publication Data
**A CIP record for this title is available from the British Library**

ISBN 1 84114 571 8
ISBN 978 1 84114 571 6

**HALSGROVE**
Halsgrove House, Lower Moor Way
Tiverton, Devon EX16 6SS
Tel: 01884 243242
Fax: 01884 243325
email: sales@halsgrove.com
website: www.halsgrove.com

Printed and bound by D'Auria Industrie Grafiche Spa, Italy

This book is dedicated to my son Bart
and his beautiful wife Charlotte,
and to the loving memory of his mother
Tessa Schneideman – a truly exceptional
painter, mime, drama teacher and
director, and friend.

Paradiso stone form, 30cm high

# acknowledgements

I would like to express my gratitude to Katharine Coleman, Jane Dorner, Candice-Elena Evans, Alexia Goethe, Sam Herman, Dan Klein, George Layton and Michael Robinson for their contributions to this book, and also to those too numerous to mention who have supported us over the years, including the many galleries in the UK and abroad with whom we have had such amicable relationships.

I would especially like to thank my colleagues at London Glassblowing, both past and present; Marie Worre Hastrup Holm and Szilvi Marks for their help with this project, and Simon Butler and Karen Binaccioni of Halsgrove – the publishers. Also Janine Christley – Ruskin Glass Centre; Stephen Fisher – Church Gallery; Lee Freed – Freed Gallery; Joanna Hayward – The World of Glass; Melanie Kidd – the Hub; Katherine Pearson – the National Glass Centre.

Profuse thanks are also due to the Workspace Group, and in particular to Maddy Carragher, for her encouragement and invaluable support. To the photographers with whom I have worked, especially Ester Segarra, Andra Nelki, Graham Diprose, Trevor Smeaton and Simon Moss. To our many collaborators, including Margaret Turner, Howard Fenn, Gayle Matthias, Peter Cannings, Chris Barlow, John Crisfield and Benson Sedgwick Limited; and also to my family, friends and faithful clients for their unstinting support.

# contents

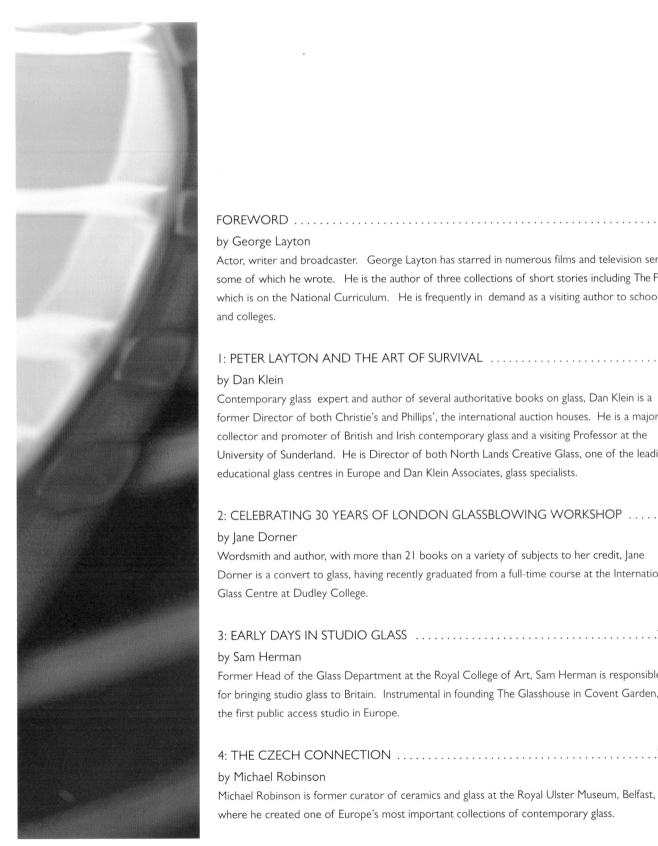

FOREWORD . . . . . . . . . . . . . . . . . . . . . . . . . . . . . . . . . . . . . . .9

by George Layton

Actor, writer and broadcaster. George Layton has starred in numerous films and television series some of which he wrote. He is the author of three collections of short stories including The Fib which is on the National Curriculum. He is frequently in demand as a visiting author to schools and colleges.

1: PETER LAYTON AND THE ART OF SURVIVAL . . . . . . . . . . . . . . . . . . . . . . . .11

by Dan Klein

Contemporary glass expert and author of several authoritative books on glass, Dan Klein is a former Director of both Christie's and Phillips', the international auction houses. He is a major collector and promoter of British and Irish contemporary glass and a visiting Professor at the University of Sunderland. He is Director of both North Lands Creative Glass, one of the leading educational glass centres in Europe and Dan Klein Associates, glass specialists.

2: CELEBRATING 30 YEARS OF LONDON GLASSBLOWING WORKSHOP . . . . .15

by Jane Dorner

Wordsmith and author, with more than 21 books on a variety of subjects to her credit, Jane Dorner is a convert to glass, having recently graduated from a full-time course at the International Glass Centre at Dudley College.

3: EARLY DAYS IN STUDIO GLASS . . . . . . . . . . . . . . . . . . . . . . . . . . . . . . .25

by Sam Herman

Former Head of the Glass Department at the Royal College of Art, Sam Herman is responsible for bringing studio glass to Britain. Instrumental in founding The Glasshouse in Covent Garden, the first public access studio in Europe.

4: THE CZECH CONNECTION . . . . . . . . . . . . . . . . . . . . . . . . . . . . . . . . . .31

by Michael Robinson

Michael Robinson is former curator of ceramics and glass at the Royal Ulster Museum, Belfast, where he created one of Europe's most important collections of contemporary glass.

5: A CATALYTIC CONTRIBUTION . . . . . . . . . . . . . . . . . . . . . . . . . . . . . . . . .39

by Katharine Coleman

One of the world's leading glass engravers, Katherine Coleman formerly lectured at Morley College and chaired the Guild of Glass Engravers. Finalist for a number of major competitions including the Jerwood, Glass Sellers and Coburg Glass Prizes, she is widely exhibited, with work in numerous private and public collections.

6: LONDON GLASSBLOWING – AN INSIDER'S VIEW . . . . . . . . . . . . . . . . . . . . . .45

by Candice-Elena Evans

Candice-Elena Evans is a freelance exhibition organiser and curator. Responsible for a number of important shows, including Wearing Glass, 9 X 9 and the British Glass Biennale, she is a former chair of the Contemporary Glass Society.

7: AN ADDICTION TO GLASS . . . . . . . . . . . . . . . . . . . . . . . . . . . . . . . . . . . . . .53

by Peter Layton

A founder member and former chair of British Artists in Glass and the Contemporary Glass Society, Peter Layton has work in many international collections. Author of Glass Art – an overview of
contemporary art glass.

8: GLASS EQUALS ART . . . . . . . . . . . . . . . . . . . . . . . . . . . . . . . . . . . . . . . . . .63

by Alexia Goethe

Specialising in modern masters and contemporary art, The Alexia Goethe Gallery has been established since 1988, in Germany, New York and now in London, with an important new gallery in Dover Street.

THE TEAM – WORK AND STATEMENTS . . . . . . . . . . . . . . . . . . . . . . . . . . . . . .69

AFTERWORD – 'THE TIMES THEY ARE A CHANGING' . . . . . . . . . . . . . . . . . . . .88

by Peter Layton

ARTISTS' CURRICULUM VITAE . . . . . . . . . . . . . . . . . . . . . . . . . . . . . . . . . . . . .90

LONDON GLASSBLOWING ASSOCIATES . . . . . . . . . . . . . . . . . . . . . . . . . . . . .96

# foreword
## BY GEORGE LAYTON

George as Puck, his first acting role, in a costume
made from our living room curtains.

When asked by some associates of London Glassblowing Workshop to write the foreword to this book celebrating 30 years of the studio I couldn't help thinking that there are scores of people far more qualified to comment on Peter Layton and his work than I, indeed that is precisely what many of the contributors to this book will be doing. Therefore, it is with unabashed pleasure and pride and without a hint of embarrassment that I am going to grab this opportunity to write more personally about my hugely talented 'big' brother.

Freddie Layton, our father, was a refugee who fled the Nazi tyranny. He came to this country in 1939 with a loving wife, Edith, Peter aged two and very little else. He fought in the British Army, survived the war, during which time I came along followed a few years later by our sister Vivienne. What Freddie Layton desperately wanted for his three children, however, was something he never enjoyed – financial security. I was a lost cause since at a precociously early age I decided to be an actor. Vivienne brought him hope when she went to college. She ultimately devoted her working life to teaching children with learning difficulties for which she was rightly honoured by the Queen, something that would have given our parents great pleasure.

Our grandparents, Dr Victor Hecht and
Rosa Hecht

Peter, on the other hand, through family contacts, got a job with good prospects in the then thriving textile industry. Good prospects were something of an understatement. Starting on the factory floor he would learn all aspects of the business with a view to ending up in the boardroom. Peter was being groomed to run the company. A job with monetary reward, security, all the trappings of 'success'. How thrilled Freddie must have been. The words job satisfaction, fulfilment, destiny were not in his vocabulary but they were in Peter's. He hated it. It must have been a dreadful time for him. What it must be like to work in a job you loathe is completely foreign to me and yet that is what many people have to settle for with lifelong regret. Peter didn't. With youthful bravery and his parents' blessing he turned his back on the 'job with good prospects' and went to art college.

9

Peter at 18 months

Two sons go into the most precarious of professions. How does this come about? Is it in the family genes? Was it the environment we grew up in? School? Not the latter, its art department was pathetic and the drama department non-existent. For Peter, there were two major influences: his maternal grandfather, Victor Hecht and the City of Bradford. This industrial town on the edge of the Yorkshire moors with its Victorian museums, art galleries and concert halls, was full of character and a wonderful place in which to grow up. It was his grandfather, a doctor by profession, an amateur painter and photographer, and lover of the arts who took him to exhibitions and concerts and nurtured the artist in Peter.

Textile's loss was our gain. The textile industry in this country is dead. If Peter had stayed in it he too would have died, if not physically, certainly spiritually. Peter's work goes from strength to strength and lives on. It will live forever. He is at the top of his tree. I may be biased but I think he is one of the greatest glass artists in the world. And the greatest of brothers, that I can say with certainty.

**George Layton**
June 2006

*Opposite:* Peter working at Roberto Neiderer's glass factory at Hergiswil, Lucerne, Switzerland, 1983

# chapter one

# PETER LAYTON AND THE ART OF SURVIVAL

Peter blowing at Morar, 1971

This year Peter Layton's 'London Glassblowing' celebrates its thirtieth anniversary, but as perhaps the first British artist to fall for the lure of free expression in the medium, when it all got going during the swinging '60s, he has earned and fully deserves a special place in the history of contemporary glass. His story is well known and well documented, most recently through a comprehensive essay about him on the Contemporary Glass Society website (see the following essay by Jane Dorner). The website also acknowledges his key role in their existence: 'The Contemporary Glass Society was established through [his] vision and commitment'. It was formed in 1997 'to represent the interests of glassmakers within the national and international community', which sums up Peter's raison d'être.

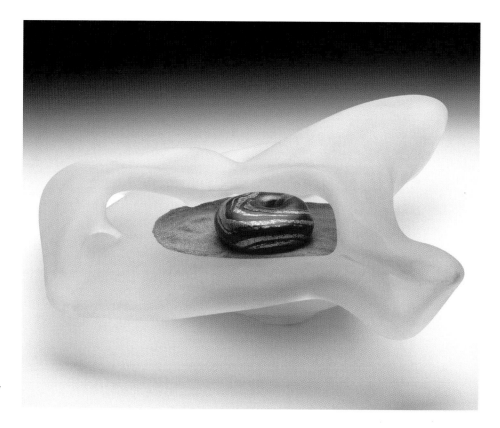

Large shell form, solid glass with nesting
iridised pebble, 1983, 40cm wide
Made with the help of Helmut Huntsdorfer

Forest Floor, large iridised funerary urn, 1984, 30cm high

Being a mover and shaker he has always wanted results and the immediacy of hot glass appealed to him. As a visiting lecturer in ceramics at the University of Iowa in 1966 Peter happened to be in the right place at the right time, right at the beginning of the American studio glass movement. He enrolled in the first glass blowing workshop at the University of Iowa in the summer of that year. It was a life changing experience and he has never looked back. On his return to Great Britain in 1968 he immediately became involved in contemporary glass here, realising its potential as untried and untested territory with enormous artistic potential, light years away from established glass traditions in Britain. Being as much entrepreneur as artist he was also interested in how it could become a sustainable way of life. Speaking recently about his partnership with Simon Moss, the designer with whom he has collaborated since 1993 on projects such as the Savoy Theatre, the Kuwait Parliament Building and various monumental sculptures for cruise liners, he commented 'I am a great believer in shared skills and collaborative effort and I enjoy the input of others in creating a synergetic result'.

Peter's talent for spotting potential and seizing the moment made him realise the importance of the Hot Glass Conference at the Royal College of Art in 1976, in retrospect a major landmark in the history of European glass. Following that, along with thirteen others, he was a founder member of British Artists in Glass, which in 1983 under his stewardship held a major conference at the RCA and an exhibition at the Commonwealth Institute Art Gallery, probably the first real survey of what was happening in contemporary British glass. B.A.G. had a sixteen-year life span until it was disbanded in 1992.

13

Since he established London Glassblowing in Rotherhithe in 1976 Peter Layton has had a policy of combining his own skills with those of fellow glassmakers, whether students, young professionals or mature practitioners. Shared responsibilities have enabled him to be where he needs to be as an ambassador of British glass at important symposia such as the 1988 Novy Bor Symposium where the scale of his work made an impression on

those who were there. It was the first indication that he would become involved in this kind of work, which has been his main livelihood for over a decade now.  It is no mean feat to sustain a studio that employs around ten glassmakers as he now does at the Leather Market premises near London Bridge, which combines a glass studio and a glass gallery.  The competition is fierce and the commercial climate as unpredictable as the English weather.  Peter knows how to make things happen whether by winning commissions, organising exhibitions, writing or networking.  He runs a welcoming and enterprising gallery as well as working in glass, and has devised a way of life that seeks to benefit other glassmakers as much as it benefits himself. Above all he is a survivor who has always managed to stay ahead of the game.

**Dan Klein**

2006

Kimono pebble form, 1986, 22cm wide
matt – acid etched
*Much of my inspiration stems from some aspect of nature or something observed while travelling. An essential element of my glass at this stage was its tactile quality.*

14

*Opposite:*  Gaudi blue vase, 1998, 30cm high,
matt – acid etched.
*Antonio Gaudi, the pre-eminent architect of the Art Noveau period, produced some of the most extraordinary buildings of all time. This series was produced as an homage, after visiting Barcelona.*

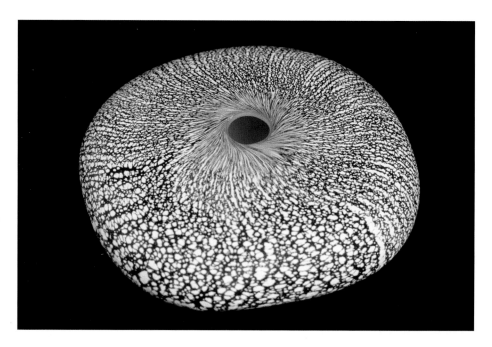

chapter two

# CELEBRATING 30 YEARS OF LONDON GLASSBLOWING WORKSHOP

Bottle, Morar, 1974, 16cm high

Peter Layton, David Johnson and David Hockney in Scotland (Edinburgh – 1955)

16

Peter Layton is a familiar figure in the glass world, not just for his work – which has always been innovative and thoughtful – but for his ebullient energy in setting up schemes and workshops, organising seminars and exhibitions, encouraging and mentoring others, and making a mark on the international glassmaking scene. There is little doubt that as a champion of studio glass, he has done more than most to advance its progress in this country.

It is 30 years since he founded the London Glassblowing Workshop (in 1976) beside the Thames at Rotherhithe (the workshop moved to its present venue in the Leather Market, near Tower Bridge, in 1995). He had already been blowing glass on and off for about a decade before that, having had the good fortune to be introduced to it in the United States just as the studio glass movement was gathering momentum. It was, he recalls, a case then of the blind leading the blind.

It was also a period of tremendous vitality and excitement: this was the onset of the swinging sixties with its promise of permissiveness brushing aside the greyness of post-war repression and gloom. The Beatles, the Rolling Stones, and Bob Dylan changed music and celebrity for ever; Mary Quant took mini-skirts (and later hotpants) way above the knee; Allen Ginsburg redefined poetry; Bridget Riley began a wave of optical illusion in art; Yuri Gagarin became the first man in space; the first James Bond film was made; and The Times eschewed its front page of Classifieds in favour of front page news. This was an exciting decade in which to be a young man on the cusp of his career, and Peter Layton made the fullest use of its opportunities.

He had known David Hockney, who was also a student at Bradford College of Art, before going on to study ceramics at the Central School of Art and Design in London, where the luminaries of the time included Ian Auld, Ruth Duckworth, Gordon Baldwin, Gillian Lowndes and others. Ceramics then was heavily influenced by abstract expressionism: leading Californian teachers such as Robert Arneson and Peter Voulkos

Clay and fused glass, made at Whitefriars Glass while a student at the Central School of Art and Design, with the help of Keith Cummings, 1965, 15cm high

having been inspired by Picasso's work in clay. This led American potters to break through previously established sculptural boundaries and reject the idea that artists working in clay should only produce utilitarian or decorative items.

In this climate, Peter Layton got his first job: a one-year visiting lecturership in ceramics at the University of Iowa. There he was swept up by the prevailing pioneering attitude, astonished by the contrast between the 9 to 5 ethic of the British, and the American students' willingness to work through the night to build equipment that might be needed for the next day. During the summer of 1966, Layton enrolled for Iowa's first glassblowing course, set up by Tom McGlauchlin who was one of the original participants in Harvey Littleton's first Toledo Glass Workshops at which the studio glass movement came into being. He was imme-diately hooked, as anyone who has followed where hot glass blows, will understand only too well.

Says Alice, ceramic head, Iowa City, 1966, 60cm high

Life in glass would have been easier for Layton had he stayed in the States, but for family reasons, he returned to England in 1968, which must have seemed at the time like going from full technicolour back to monochrome. In America, he had been riding on the crest of a wave: returning to England meant that his own work evolved in fits and starts and he was largely self taught. It was a time of recession and though the phrase 'swinging sixties' originated here, that didn't impinge on glass art. It was the year in which theatrical censorship was finally abolished, but the same freedom of expression in glass was light years away – perhaps it hasn't yet arrived. The concept of 'funk art' in glass, as in ceramics – of non-functional objects that expressed the qualities of the medium and of light and space – was alien in this country. Peter Layton had to create his own following as well as persuade art colleges such as John Cass, Croydon, Medway and Hornsey Art Schools to sit up and take note of glass – indeed he started the glass department at Hornsey and it ran for about a decade before closing down.

Group of iridised pieces, including work by
Peter Layton, Siddy Langley and Carin Von
Drehle, 1980

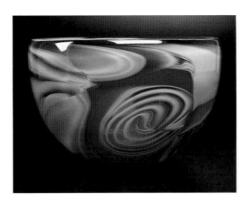

Chalcedony thick bowl, 2003, 15cm high

Peter blowing a large dish at Rotherhithe, 1984

Floral, 1992, 16cm high – acid etched

*The floral pieces blossomed around the time of one of the Monet exhibitions. Achieving the delicacy and translucence of watercolour in glass has made this a most successful series.*

Peter Layton was not the only protagonist of glass in the UK. Sam Herman, one of the graduates from the first glass course in the USA (at the University of Wisconsin), came to London to head the Glass Department at the Royal College of Art. Two years later in 1969 he conceived and set up the Glass House in Covent Garden, initially as a working outlet for RCA graduates and later as a public access studio and gallery. Peter Layton helped build the first furnace there and was involved in its beginnings. The experience had a profound impact on the style of his own work.

'The glass in the gallery always looked marvellous because the back-lighting there made it appear magical. It drew people in. But other galleries didn't follow suit, because they lacked specialist lighting and thought their clients wouldn't have it either. Encouraged by Pan Henry of the Casson Gallery I experimented, trying to find a way in which the glass would glow without fancy lighting. I came up with the iridised colours for which my studio first became known.'

Layton's blown work has since moved on. He has always favoured organic tactile forms that curve within a controlled asymmetry. Now his pieces deploy a painterly approach with shapes that emphasise their striking colours and surface patterns. Some, like his Chalcedony Series, swirl the coloured glass to make patterns like rock strata or quartz layers. Another, the popular Floral Series, appears to have drawn its watercolour effect

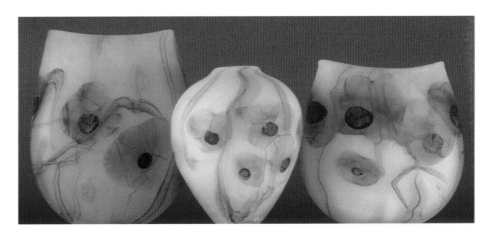

Reef stone form, 2004, 25cm high

*The work is constantly evolving and changing
and Reef has grown out of a desert inspired
series called 'Sahara'. The pieces have become
much more colourful with an underwater
character which evokes the Barrier Reef,
one of my favourite places in the world.*

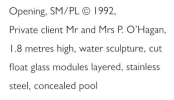

Opening, SM/PL © 1992,
Private client Mr and Mrs P. O'Hagan,
1.8 metres high, water sculpture, cut
float glass modules layered, stainless
steel, concealed pool

Sunburst, SM/PL © 2002,
Celebrity Cruises (RCL), recessed wall
sculpture, 2.25 metres high x 1.34 metres
wide, layered float glass within a polished
stainless steel void, concealed colour
changing edge emitting fibre optic lighting

from Monet, although he says it was actually inspired by the work of another painter, whose work he owns, Lilliane Delavoras.

Latterly, the blowing studio has been partly supported by architectural commissions combining glass with metals, that Peter has been working on in collaboration with Simon Moss since 1993. Moss describes their work, 'Our contrasting and complementary approaches to problem-solving mean that we are always questioning each other's ideas. We use glass and mirror-polished stainless steel to bring out visual ambiguities. Our exterior pieces react to the changes in natural light, although we often incorporate integral lighting to reveal further hidden layers within the work.' Many of these were made for luxury cruise ships, but world terrorism has seriously impacted on this side of the tourism market. Nevertheless, the pair still get short-listed for prestigious architectural projects. Peter admits that this is an important aspect of his collaborative work because making a living out of blown glass is difficult in this country. 'I'd really like to be doing more sculptural work now, and using it to express the concerns I have about what we are doing to the planet.'

This sounds like a return to a younger self from a man who is approaching a birthday with a big 0 in it. His first major solo exhibition in 1972 showed mixed media installations that amusingly pin-pointed society's drive towards disintegration. More recently, he has shown *Battery,* one of a series entitled *Matters of the Heart,* a caged assembly of

Strange Fruit, 2003, 60 cm diameter
*Strange Fruit is part of a series entitled 'Matters of the Heart', and is inspired by the invitation to participate in an exhibition called 'The Tree Of Life'. Never before have we had so much power (for good or evil). We can now manipulate human genetic development and thereby challenge the very nature of life and society. Seldom have the dualities of living been in starker contrast, nor has the potential for global conflict and its attendant horrors been greater.*

36 blown glass hearts, 21 felt jackets and fibre optic lighting. He says of it, 'Between birth and death there are only precious present moments, encompassing love, joy, passion, strength, courage, wisdom, humour, generosity, creativity, freedom – and, of course, their opposites. In squandering these we are obliged to live with the consequences.'

*Opposite:* Axios, SM/PL © 2004, Suspended rotating atrium sculpture for Prudential Assurance, 8.5 metres high, mirror polished stainless steel, positioned over reflecting water feature

So what has Peter Layton accomplished in his 30 years of running a studio and 40 years as a glassmaker? His life has been a success: his work is known from books and high profile exhibitions all over the world. His own book *Glass Art* (A&C Black, 1996) is an informed survey of the studio glass movement. He has been an astute businessman although he himself would argue this point, and he is in fact surprisingly modest. In conversation he fails to mention being awarded an Honorary Doctorate of Letters by the University of Bradford in 2003 for his contributions to the development of British and International Studio Glass.

Please keep off the grass, chromium plated plastic, concrete and metal, 1972, 90cm high, 185cm x 185cm wide

23

Six pack, silk screened ceramic and flocked metal, 1972, 35cm high – Commonwealth Art Gallery 1972

Is he perhaps saddened that despite his, and others', best efforts, Britain still does not have a well developed market, nor many dedicated collectors of contemporary glass art? It must be wearisome to continue to energetically cultivate and challenge the public's perception of glass. Those who follow in Peter Layton's footsteps owe him a debt of gratitude for all the work he has done in paving their way.

**Jane Dorner**
2006

Encapsulated, 2006, 20cm high – Uranium glass, glass, metal, rubber

*Opposite:* Flotsam and Jetsam iridised bottle, 1982, 24cm high
*Inspired by litter floating on the Thames, as seen from the balcony of the Rotherhithe workshop*

Small vase, Morar, 1974, 10cm high

Like many things in life one does not realise how well or how dramatically certain experiences will shape one's future when actually doing them. My early involvement in studio glass was certainly one of these.

I was very fortunate to be one of Harvey Littleton's first students and to be involved with him in shaping the studio glass movement. It is difficult to convey in words the camaraderie and good will that prevailed in the glass area at the University of Wisconsin. People may not realise that none of us, including Harvey, knew very much about glass blowing. Even simple things like learning how to put on a punty had to be discovered by trial and error. Consequently it was a case of learning from each other, which created an atmosphere of enthusiasm, good will, and friendships that have lasted over many years.

Vase, Morar, 1974, 17.5cm high

To give some idea of how things were at that time, we had just moved into a Quonset (Nissen) hut and because of everyone's enthusiasm the furnaces were worked 24 hours a day. Each student was given an allotted time to work with the glass. Once a week we met and all the work that had been produced throughout the past week was put on display and questions were asked about how certain effects were achieved. Often whoever had discovered some technique would demonstrate it to the rest of the class. Naturally this created a unity of sharing.

The studio glass movement was only possible because of the driving force of Harvey Littleton and the technical input of Dominic Labino. Littleton was a potter whose father worked at Corning Glass and obviously this aroused an interest in glass in him. Dominic Labino was a research scientist in glass who, by creating a small tank furnace that was affordable, made it feasible for individuals to work the glass in an artistic way. This all came about in two symposiums organised by Harvey in Toledo, to which he invited individuals who might be able to contribute knowledge on the possibilities of working glass on an individual basis. One has to remember that at that time there was a general

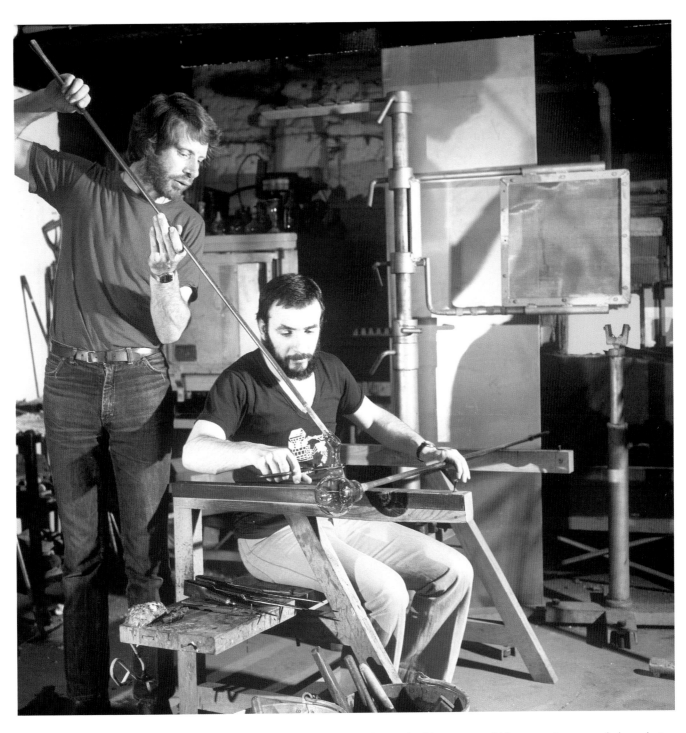

Peter and Norman Stuart Clarke in the studio at Rotherhithe, 1979

attitude that glass could only be worked by a team of blowers using not only large but expensive furnaces. Thanks to the efforts of Labino and Littleton and many others it became possible to find a viable way of working glass in much the same way as studio pottery was being produced at that time.

I remember once when we were invited to visit Dominic Labino's studio, and one of the students was making a piece of glass to show Dominic a particular technique that he

had developed. I was standing next to Dominic and Harvey and as the piece progressed and he was about to place it on the punty, it fell off and skittered across the floor. He quickly gathered a bit of molten glass out of the furnace and picked the piece up off the floor. Harvey very proudly turned to Dominic and said 'see, I teach my boys how to pick them up when they drop them'. Dominic without batting an eyelid said to Harvey 'why don't you teach them not to drop them?' I should hasten to add that there was nothing that we could teach Dominic, and in fact until then we knew very little about colour, but he had developed a whole range of colours that were compatible with the glass we were using and was extremely generous in sharing his knowledge with us.

There are many incidents that I could cite showing the generosity and openness and camaraderie that existed in the early days of studio glass amongst all the people involved. We were all learning and were glad to share our newfound knowledge with each other. I consider myself very fortunate that because of luck and circumstances I was able to be a part of it.

28

Since then almost fifty years have passed, and what incredible changes have occurred in that time, and what an enormous influence studio glass has had on the glassmaking industry. I am very impressed with the current contemporary glass scene and that glassmakers like Stephen Newell, Anthony Stern and Peter Layton have managed to maintain successful studios for so long, providing inspiration and training for the new generations of glass enthusiasts. I can only congratulate and wish them well.

To conclude, I remember when I was working late once, around 2.00am in the glass studio at Wisconsin. As the studio was in the downtown part of Madison, people would

Paradiso perfume bottle, 2004, 15cm high

Ariel stone form, 2004, 20cm high

often walk in. On this particular night a very drunken couple came in just as I was re-heating a piece before placing it in the annealing kiln. As I was putting my pipe away the woman lurched rather dangerously towards the opening of the furnace, which naturally alarmed me. I quickly stepped between her and the furnace. She then turned to me and asked, 'How are you fishing those things out of there?' Somehow in her drunken state she thought that the glass pieces were ready made in the furnace and that I was just fishing them out. Well in a funny way that is exactly what we glassblowers have been doing all these years – somehow, hopefully, 'fishing' a creative expression of ourselves out of the molten glass.

**Sam Herman**

2006

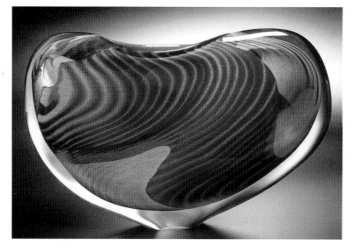

30

Paradiso stone form, 2004, 20cm high

*Paradiso represents a more painterly approach to vessel making in which flattened forms become the canvas for strong brushstrokes of vivid colour. Paradiso evolved during months of careful colour experimentation to achieve warm and vibrant patterns, evocative of Eden.*

*Opposite:* Pyramid, 1988, 250cm high, hot cast glass bars –blown central column by René Roubicek and Petr Novotny. Collection Lemberg Castle, CS

31

By the mid 1970's studio glass that furnace-manipulated phenomenon that Sam Herman had introduced from the USA in the late 60's, was an established and increasingly apparent feature of the British crafts scene. Glass departments were spreading in universities and an evangelically enthusiastic and expanding congregation of hot glass adepts poured out their offerings in craft shops and galleries and wherever they could find exhibition space. The 70's was a period of great enthusiasm and exciting adventures in all sorts of mediums but the glass thing happened so quickly it seems in retrospect almost like spontaneous combustion. Heads of departments and lecturers in colleges were only a degree's duration beyond the students they were teaching and they all shared in the physical, intellectual and spiritually liberating adventure of exploring an entirely new medium that had no discovered horizons or boundaries, no limits beyond those imposed by one's own skill and vision and no taboos or restrictions imposed by history, criticism or theory. Anything went.

The whole thing spread like a tide from America to Britain, Europe and then on to Australia, New Zealand and Japan. Obviously older glass cultures like Venice, Germany and Scandinavia and individual artists and designers contributed to the phenomenon but it all seemed like an essentially American thing, exuberant, spectacular and driven by the larger than life virtuoso performances of its superstars and their incredible exploits.

Then the Czechs arrived. It wasn't as if they appeared overnight. They were the second oldest glass culture in the world, but locked away behind the iron curtain they were just not part of the razzmatazz that typified the studio glass movement. When René Roubicek produced his monumental installation at the Brussels Expo in 1958, there had been no studio glass artists to see it and most of the glass industries were too comatose or preoccupied with their own problems to notice. In world fairs through-out the 60's Czech glass artists, trained and matured in their own incomparable educational system, had been winning accolades for their individual contributions as well as for the designs that had made the Czech glass industry so successful. Pavel Hlava had

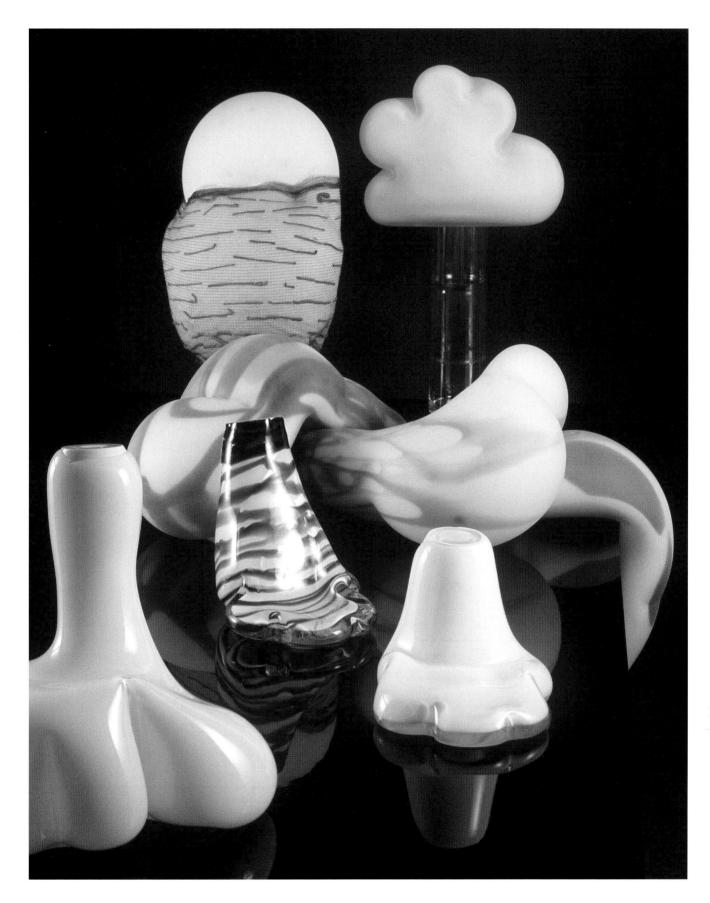

Hurdle, a glass tank-trap, 1985, 200cm high.  Hot cast glass, metal
A subversive comment on the Russian occupation of Czechoslovakia.
Collection Lemberg Castle, CS

spent time at the Royal College of Art in the mid 60's, too early to have an influence. His show at Rosenthal's London showroom in 1973 was an inspirational demonstration of idiomatic hot glass usage but unfortunately seen by too few to make the impact it merited.

It therefore came as a revelation when Professor Stanislav Libensky presented Czech glass to the international conference 'Working With Hot Glass' at the Royal College of Art in 1976. All the stars were there, a hot glass rodeo with broncos bucking in every direction, and then Libensky's presentation showed clearly what reason, maturity and total mastery could do to express controlled, deep, emotional, social and spiritual state-ments.  Glass as art rather than super craft.  Like the films of Andrei Tarkovsky and the science fiction of Stanislaw Lem – it was an eye opener: art from a totalitarian state that shocked through its passion and intensity as well as its performance.

Few people at that conference could fail to be affected by Libensky's quiet, dignified charisma and the power of the work he had shown, Peter Layton was no exception. He had just opened his London studio and invited the Czechs to visit his first solo exhibition at the Camden Arts Centre. Born in Prague where at the time, his father was

Peter Layton, René Roubicek, Finn Lynggaard,
Joel Philip Myers, Willem Heesen, Marvin Lipofsky,
Novy Bor, 1988

Peter casting the bars for Pyramid,
Novy Bor, 1988

a manager in a Czech glass factory, Layton studied ceramics at the Central School of Art and Design, London. His conversion to glass was inspired by his experience of the American studio scene, whilst teaching there. He was self taught as a glassmaker and conscious of his limitations, so unstinting praise and encouragement by masters like Libensky, Marion Karel and Ales Vasicek was enormously gratifying and stimulating. The contact did not end there, certainly not for Layton and a number of other British glass-makers. In 1982 the Czechs opened their vast Crystalex glass factory in Novy Bor to an international working conference, which attracted big names from across the glass world. This was followed by even bigger events in Novy Bor in '85 and '88 with massive international participation. Foreign glassmakers were given access to a scale of

Endless Column SM/PL © 1991,
Suspended sculpture for the exhibition
'Le Verre' Rouen, France, 8 metres high

working that they could barely imagine in their own small studios: factory size furnaces, equipment and materials, teams of highly trained and experienced masters as assistants and the stimulating contact with Czech artists and designers.

For Layton, these Novy Bor symposiums were a crowning achievement. Able to work on a massive scale, he used the opportunity to produce sculpture he couldn't have made at home. In 1988, if these had been the Glass Olympics (as they were jokingly referred to) he would, according to Libensky, have been awarded a gold medal for his enormous Pyramid built from separately hot cast (poured) glass bars, displayed in a show that included Dale Chihuly, Billy Morris, Joel P. Myers, Marvin Lipofsky, as well as the hosting Czechs. It was not only a great achievement for a British glassmaker abroad but it also won for him considerable influence and standing in Europe, particularly amongst younger emerging artists, many of whom would soon enter a post communist art world.

The Czech influence on the world studio glass movement may not have been recognised before 1976, despite their spectacular show at the Montreal Expo in 1968, but today it is one of the foundations of the movement. A number of artists like Stephen Proctor, Ronald Pennel, Diana Hobson, Colin Reid and Peter Layton had the opportunity to visit Czechoslovakia and experience its working methodology under communism. Since the velvet revolution it is becoming easier to study or work there for prolonged periods and artists like Angela Thwaites and Hannah Kippax have done so.

As more and more Czech and Slovakian artists exhibit abroad their work becomes a more familiar element in the art world. The Studio Glass Gallery in London has played a major role in its dissemination. Today Czech and Slovakian artists are themselves familiar figures as conference delegates, exhibitors, visiting lecturers and professors in British Universities. Sunderland University's glass department has had resident professors e.g. Dr Sylva Petrova and Zora Palova, visiting professors such as Libensky when he

Gate of Dreams SM/PL © 1997, RCL,
water sculpture, 2 metres high, pierced
laser cut stainless steel, polished Altuglas,
aluminium finial, marble base

Sentinel 1/2/3/4 SM/PL ©2006,
Private commission for MY Talisman C,
niche pieces, 0.9 metres high, kiln cast glass,
stainless steel

*Opposite:* Aurora flared vase, 2004, 15cm high
*Sometimes things just click, and a new series like
Aurora emerges through 'sketching on the blowing iron'.*

was alive, and the world famous engraver Jiri Harcuba. Some aspects of Czech glass, particularly large scale casting, and optical cutting and polishing techniques are obviously visible in the work of many important international artists. Because of the immense diversity of the routes taken in glass by Czech and Slovakian artists, as different as Libensky, Kopecky, Cigler, Harcuba, Zemecnikova, and so on, such individuals frequently become role models for students, either long term or only briefly. For many impressionable students they become ingredients in that melting pot which is their background. Whether they will be conscious of this influence in their own finished work is debatable, but its contribution is undeniable.

Peter Layton is one of the many who have benefited from the Czech connection. A pioneering figure in British studio glass his work has always tended to be confidently and at times dramatically decorative. The Czechs gave him the opportunity and the inspiration to explore a scale and intensity he might never have experienced otherwise and whose effect has remained a corner stone of his vision in glass.

**Michael Robinson**
2006

Crystal Stream SM/PL © 2002,
Celebrity Cruises (RCL), pool head exterior
water sculpture, 4.5 metres wide, mould
blown glasss, stainless steel trays, internal
integrated fibre optic lighting

Metamorphosis, 2004, 25cm high

Peter Layton, one of the pioneers of studio glass in Britain, is not only still producing new and original work, he has also fostered and encouraged more young aspiring artists in his London Glassblowing Workshop than any other studio. Generous to a fault with his time and resources; he is acknowledged by many for both his kindness to others and the significance of his contribution to the studio glass movement.

Born in Prague and brought up in England, Peter Layton studied ceramics at the Central School of Art and Design in London under some of the foremost potters of the early 1960s. While teaching ceramics at the University of Iowa he encountered the new phenomenon of studio glass, those first attempts at making glass outside the factory environment, and for him it was a *coup de foudre (a 'bolt of lightning', the phrase describing instantly falling in love*) from which it appears he has never recovered. The alchemy of glass, the process of making and forming it, its myriad properties and colours all appealed instantly and ineradicably. There were few formal techniques to be learned at that stage. Making small furnaces and discovering new forms of glass became

Paradiso dropper, 2005, 29cm high

Ebb Tide tall stone form, 1999, 39cm high

*Opposite:* Paradiso turquoise stone form,
2005, 46cm high

part of the steep learning curve that so many glassmakers of his generation experienced.  That initial excitement and fascination is still apparent in his work some forty
years later. Peter's gentle and generous approach to promoting studio glass and encouraging others may well reflect his experience of the very best of America at that time.  It
was a time when anything seemed possible, an attitude that allowed for the occasional
disaster, which took setbacks in its stride and made something positive of them. The
ethos of sociability and cooperation was definitively a crucial part of the mix.

In the early 70's Peter Layton was instrumental together with Sam Herman, in setting
up the Glasshouse in Covent Garden.  He subsequently established his own small glass
studio at his pottery at Morar in the Highlands of Scotland, a Glass Department at
Hornsey College of Art (Middlesex University) and in 1976, the London Glassblowing
Workshop on the Thames at Rotherhithe.  Moving to the Leather Market site in 1995,
this colony of young makers has continued to grow with Peter at the helm, changing and
progressing alongside the individual careers of so many glass artists fostered in both the
hot shop, the gallery and the cold shop.  Making series production as well as major art
pieces, whether for sale in galleries or as commissioned installations on ocean liners and
in new buildings, the members of London Glassblowing, under Peter's light touch, take
on an increasing range of challenges.

His own work has always remained in the spotlight when contemporary glass is
mentioned.  Among the first exhibitors at Chelsea Crafts Fair, Peter has exhibited in
almost all the galleries that show glass in Britain, and his work is also held in public
collections in the UK, the Czech Republic, the USA, Germany, Japan, Switzerland and so on.

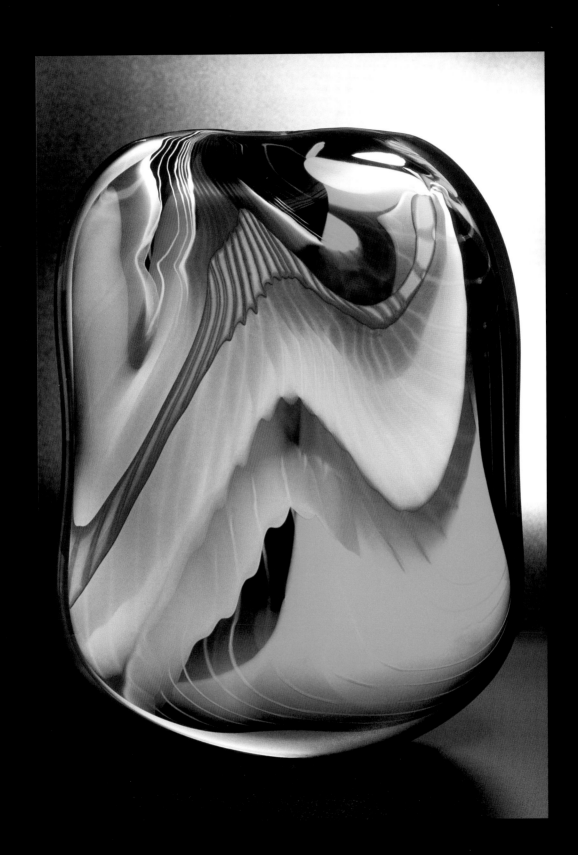

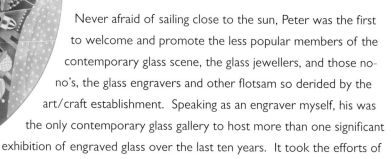

Circus, double cased dish, sand etched, made in collaboration with Gayle Matthias, 1994, 33cm diameter

Never afraid of sailing close to the sun, Peter was the first to welcome and promote the less popular members of the contemporary glass scene, the glass jewellers, and those no-no's, the glass engravers and other flotsam so derided by the art/craft establishment. Speaking as an engraver myself, his was the only contemporary glass gallery to host more than one significant exhibition of engraved glass over the last ten years. It took the efforts of Peter, and especially Professor Dan Klein, to bring serious public attention to the possibilities of cold working as a form of contemporary glass art. Anthony Scala's (recent winner of the 2005 Glass Sellers Prize) success is patently due to Peter's active encouragement.

In 1997 it took Peter's drive and energy to initiate the formation of the Contemporary Glass Society from the cold ashes of British Artists in Glass, introducing a wider range of conference and workshop activities which now bear witness to the flourishing society of today. Always an enthusiastic participator, Peter was invited to numerous symposia and conferences as a guest artist whose work surprised even the great Czechs themselves at the Novy Bor Triennials in the 80's.

Within the Worshipful Company of Glass Sellers, of which he is a Freeman, he has also wrought his characteristic magic, conjuring up from the traditional ceremony of a handshake and dinner for the winner of the annual Glass Sellers Award, an exhibition in 2005 of the work of all the previously unseen finalists, and guest exhibitor Peter Dreiser. Imagination and a talent for empowering others are what lie within this magic. Certainly his role as a catalyst has produced very positive results and the burgeoning contemporary glass scene would be a far poorer place without his continuing involvement.

**Katharine Coleman**

2006

*Opposite:* Nimbus, 1989, 26cm high – acid etched

# chapter six

# LONDON GLASSBLOWING – AN INSIDER'S VIEW

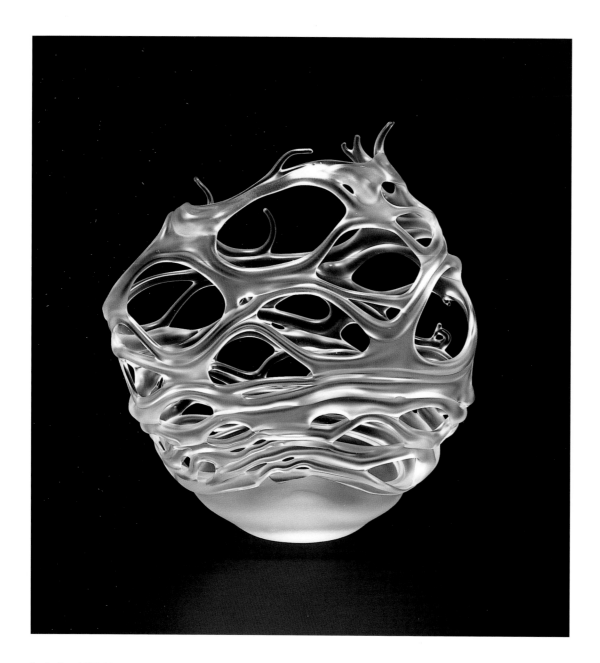

Ice Basket, 1997, 18cm high, matt –
acid etched

*The Ice Baskets seek to express the magic of glass, its sensuality and fluidity. They explore the theme of ice and snow, exploiting the way glass freezes at a particular moment in the cooling process. Such pieces record intention and accident, a process partly controlled and in part natural, in the effort to create objects that express more than purely functional or decorative qualities.*

There are markers in one's life; turning points that change it forever sending you in totally unexpected directions. I was looking for my own artist's studio somewhere near the Thames in Rotherhithe one day in 1995, when I stumbled upon London Glassblowing Workshop (LGW) and met the delightfully chatty Victor Ramsay, who within the hour had persuaded me to contact Peter Layton at a later date. Since I had just graduated with a degree in Fine Art, Sculpture, and had previously used glass in some installation pieces for my final show, my interest was aroused but I was nervous to make the call. Having plucked up courage, I was greeted by Peter's friendly voice inviting me to visit the studio again. During a tour around the hot shop, he told me that the studio would soon be moving to the Leather Market near London Bridge, and a few months later I joined his crew as an assistant glass finisher, with the opportunity

Wave SM/PL © 1995, RCL,
water sculpture, 1.8 metres high,
stainless steel and brass fabrication,
float glass, integral fibre optic lighting

Ariel V-form, 2004, 35cm high
*Ariel is a near relative of Chalcedony and stems from requests to make a version that emphasises the blues.*

to continue with my own work and experiment using his kiln. This was my first real encounter with glassblowing, and the team, which at that time included: Patrick Stern, Martin Andrews, Tom Petit, Tim Waldegrave and Layne Rowe; Lesley Sholes and Victor Ramsay in the Glass Art Gallery; and the architectural glass designer Simon Moss.

Having always been enthusiastic about the arts, the creative process and its attendant exchange of ideas, I found that the glass community, albeit small, was welcoming and supportive of this. Glass was certainly a great path to follow and the glass world a great field to have entered. Through hands on experience, I acquired the skills to complete the studio pieces before they were taken to the gallery. These included the basics of grinding, finishing and acid etching. I was also lucky enough to be able to assist Simon, who at that time was working on the very first large-scale sculptural commission for the Royal Caribbean cruise ship, 'Legend of the Seas'.

With my background in Fine Art, I hardly expected to catch the glassblowing bug but I began having evening classes with Martin, and eventually took on the role of a glassblowers' assistant. This allowed me to experience an entirely new creative process, and become familiar with the challenges posed by the immediacy of the material, and the spontaneity of decision required throughout the making process.

47

Never a dull moment! Every day, whether working on series pieces or experimental works, there was always something new to learn. Being part of the team, the physical work, the challenge of a new design, and everything else that makes London Glassblowing what it is, captured my enthusiasm and kept me there for eight inspiring years.

Spirale dropper, 2006, 20cm high

*Opposite:* Janus SM/PL ©1995, RCL, Overhead atrium sculpture, 9 metres high x 15 metres wide, stainless steel, brass, toughened float glass with fibre optic edge lighting

As an assistant one was guided, yet encouraged to use one's own initiative and creativity, and this included setting up the gallery for exhibitions, attending fairs, organising photography, dealing with customers at open weekends, and building relationships with those clients whose faces became familiar. Peter usually invited the studio team to be part of the shows in the gallery and sometimes extended this invitation to include us in exhibitions with other galleries. These group shows provided goals, a chance to develop and complete work and present it to the public.

What ultimately lies at the heart of London Glassblowing?

The studio is an arena, a place where what you make of it, is what it is. What you choose to put into it can be incredibly rewarding, providing opportunities to take responsibility, make choices and explore possibilities within the resources available. The energy generated through cooperation and collaboration was and is unusually positive. It is based on an ethos that allows for a shared creative freedom, a synergistic

Spirale thick bowl, 2006, 13cm high

Kimono coral form, 1989, 20cm high –
acid etched

experience that invites creative input and involvement of which the outcome is the
consistently excellent quality of the work produced.

Over the years London Glassblowing has seen a steady flow of artist makers come and
go, some pass through, while others stay much longer.  Some even leave and return,
like Layne, Marie and myself.  I have fond memories of the studio and am delighted to
remain in contact with several of the old team.  Working there, long-term friendships
have been created, as were the foundations of my independent projects like Wearing
Glass 2006 and it was there I was first introduced to the Contemporary Glass Society
(CGS), when Peter established it after the demise of British Artists in Glass (BAG).
CGS events provided occasions for us to meet other glassmakers outside the studio
environment.  Eventually I became a committee member and then Chairman, which
in turn led to my links with the Ruskin Glass Centre and my curating the British Glass
Biennale.

Peter has established a reputation not only as a glassmaker, but also as a tutor, mentor
and expert in his field.  He has created a very accessible and open studio/gallery,
encouraging artists, and glass graduates to take work placements in the studio, helping
them to get started.  He has always been supportive, critical and fair.

*Opposite:*  Kimono coral form, 1989, 30cm
high – acid etched

People are key to the success of any venture, and personalities, skills, hard work and a passion for glass are important to the survival of any studio. This also requires the sustained interest and support of the public, be it Peter's longstanding customers or new collectors and enthusiasts. I am convinced that the studio's constant evolution and consistent creativity will ensure that it continues to thrive.

I am delighted to have been a part of it!

**Candice-Elena Evans**
2006

52

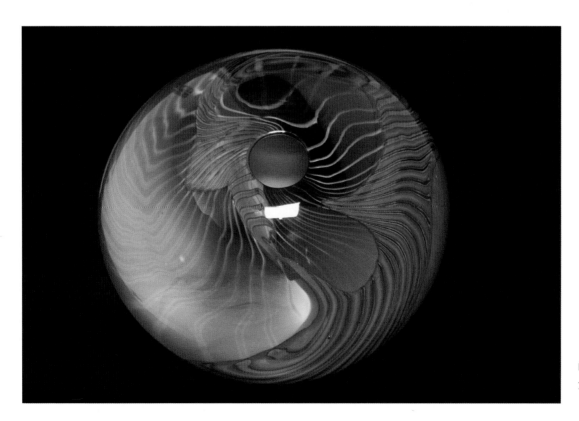

Paradiso pink sphere,
2005, 30cm diameter

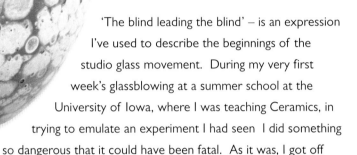

Flotsam and Jetsam iridised tall ovoid,
1984, 24cm high

'The blind leading the blind' – is an expression
I've used to describe the beginnings of the
studio glass movement.  During my very first
week's glassblowing at a summer school at the
University of Iowa, where I was teaching Ceramics, in
trying to emulate an experiment I had seen  I did something
so dangerous that it could have been fatal.  As it was, I got off
lightly with a right hand so badly burned that it almost put paid to a promising career
in glass.  But this medium is addictive and so insidious that it literally seduces people.
Not just the beautiful sparkly stuff itself, but more especially the process of working the
molten 'metal'.  Once you've tried it, you're hooked!  That's my experience, and that of
thousands of other glass addicts.  For years, I resisted any thoughts of moving from
ceramics into glassmaking but eventually I couldn't hold out any longer and in 1976, with
Victor Ramsay's help I set up the London Glassblowing Workshop in an old towage
works, overlooking the Thames on the south bank, between three famous pubs, the
Angel, The Mayflower and the Ship.  We oscillated between them.  Actually that isn't
true, for though we visited them all from time to time, mostly we didn't even stop for
lunch because the furnace was blazing away and the gas bill, not to mention the rent
and so forth, had to be met.

In the beginning it was a part-time operation, all done on a shoestring.  There was no
working capital other than my personal overdraft.  There was no market for studio glass
and we had little or no experience of running a hot shop.  At first we switched our
tank furnace on and off each week to save money while I was off teaching ceramics
to subsidise the studio. Victor, whose previous career had been working out a system
to make his fortune from horseracing, was also a part-time house husband.  We had
met while learning to relax at the National Childbirth Trust, along with our respective
wives; in other words we slept while they huffed, and having that in common seemed
to provide a positive basis for working together and so we did, for many years.

Lazuli coral form, 1997, 38cm wide

*Lazuli is a series that developed in response to a challenging commission. The brief was to make a table centrepiece whose colours represent the sun, sea and sand.*

As a beachcomber from way back I attempted to capture the sun, the sea, sand and sky in my work often inspired by shellforms, pebbles and lichens. Themes like ice and snow have also preoccupied me and I have sought to exploit the way molten glass cools and freezes at a certain point as it is manipulated. A frozen moment that captures and expresses a gestural statement of accident and intent recording a process in part natural, in part controlled, in the hope of transcending purely functional and decorative qualities.

After a while we were joined by Norman Stuart Clarke, a graduate of the Glass Department of Middlesex University (Hornsey College of Art), which I had started some years earlier. He also had a year or so working in industry (at Nazeing Glassworks) where he'd learnt a thing or two. In any case he knew more than we did which made him a valuable addition to the crew. Both Victor and Norman are great characters, full of risqué tales, homespun philosophy and trivia that kept us royally entertained before, during and between pieces. Having decided that we needed a type of glassware that looked good in any lighting, we researched iridising together eventually creating an amazing palette of colours. What a eureka moment, opening the annealer to discover for the first time that we'd achieved the most incredible lustrous electric blues – could we repeat them? We did, and then moved on to gold and silver as well as more delicate pearlescent effects.

Volcanic series iridised shell form, with silver work by Howard Fenn, 1984, 20cm wide. Collection Ebeltoft Glass Museum, Denmark

Next to join us was Siddy Langley, a recruit from the week-end classes that I gave. She soon sold up her clothing boutiques, staying for nine years or so before starting her own studio, Alchemy Glass. I believe Norman and Siddy still make iridised work amongst other things, but it's an unhealthy business; very unpleasant fumes which can have serious side effects. Some years ago I heard of a Hungarian glassmaker who died from iridising carelessly.

Other people came and went, I gave up teaching ceramics, time flashed by. The work got better and sales improved. Simon Moss joined us and he and I began to work together on large sculptural and architectural projects, which was fortunate since the bottom dropped out the retail market during the recession in the 90s. For several years we found a niche in the cruise line industry – combining glass with metal to create large and complex sculptures on board the new generation of super cruise ships. This came to an abrupt end following the unfortunate events of 9/11. Years later, we are still collaborating, attempting new land-based works on an even grander scale, some of which no longer include glass. Our latest proposal for which we have been short listed in a competition to design a sculptural Gateway to Yorkshire, near the Humber Bridge, is a 30 metre high stainless steel and glass structure entitled The Yorkshire Rose. We have also recently begun to work together on a series of kilncast forms – a new and exciting direction for us, demanding new skills and a different way of thinking.

Over the years many interesting people have worked, trained and collaborated with me in the studio. Several have gone on to set up their own workshops – Siddy in Devon, Norman originally in Cornwall, and now in France, Victor in Scotland and others elsewhere.

Gateway to Yorkshire night shot, competition
short-listed proposal, SM/PL ©2006,27 metres high,
Steel, stainless steel, float glass, lighting

*Right:* Photomantage visual for Cass Sculpture Foundation to accompany commissioned silver maquette, SM/PL © 2006

*Below:* Viewfinder SM/PL © 2005 Gateway sculpture proposal visual for Bradford regeneration scheme, 8.8 metres high, bright polished stainless steel

*Opposite:* Waveney SM/PL © 2005 Landmark sculpture (meeting place) proposal visual for town centre regeneration project, 5.8 metres high

59

Lens SM/PL ©2002,
Celebrity Cruises (RCL),
kiln cast glass panel,
120mm diameter, dichroic
glass interlayer

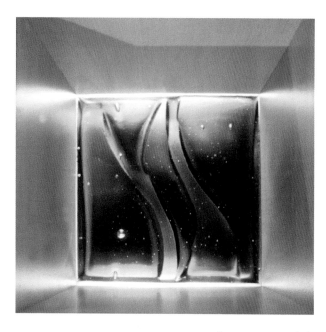

Relief panels SM/PL ©2002,
Celebrity Cruises (RCL), fused then
slumped layered body tint glasses, 0.2
metres x 0.2 metres

60

Some now work for other people, and others have moved into related fields like curating, while yet others have ultimately moved on to other things although they reappear periodically – glass has that effect on people– it's a love affair that you never really get over!

I've been extremely lucky to have had the opportunity to work with such incredible people, all of whom, without exception, have made a worthwhile and in some cases, a very significant contribution to the studio. Their enthusiasm and hard work are what have made the ups more enjoyable and the downs more bearable and each has brought something fresh and valuable to the mix.

Balance series, SM/PL ©2006,
35cm high, kiln cast glass, two
intersecting glass elements,
polished and satin finishes

I've never had a better team than at present and they are the friends I refer to in the title. I hasten to add that I do have more than ten friends, but these are the people I spend time with day after day, and although there are occasional problems and disagreements, the underlying synergy derives from our common aim and compulsion to explore this wonderfully versatile material, discover new possibilities and produce the best glass we can at any given moment. Sometimes our work is bright and colourful, at other times it's subtle and calm, sometimes it's more concerned with form, and in yet another phase with surface qualities. At best, these all come together and occasionally truly exceptional pieces emerge. If these are rare, then the pursuit to achieve a form of perfection– in the sense that each aspect works – is what keeps us going – like always searching for that perfect pebble on the beach.

**Peter Layton**
2006

Libra I, SM/PL ©2006,
33cm high, kiln cast glass, two intersecting
glass elements, polished

Personne ne joue, ne joue plus qu'un artiste ;

Qu'un artiste qui est engagé sur une route parallèle/perpendiculaire.

- A' quoi?

- De rien!

- Le verre : une dose de silicium, de feu et du souffle.

- Et puis, quoi?

Mère Nature offre ; les artistes composent dans un vaste champ d'activités. Il se trouve que Peter Layton fait surgir de tout cet amalgame physique des créations qui sont les résultantes de son imaginaire créatif.

Peter = le verre. Ca se voit, ça se comprend!

*Opposite:* The Game of Life, Win – Win,
Lose – Lose, 2006, 60cm diameter –
Plastic, glass
*This installation comments on society's drive
towards disintegration, and the precarious
nature of life. The text reads: In the moment
of remembering the millions of innocents
slaughtered in the name of religion, ideology
or for no other reason than that they were
supposedly different from those who killed
them, one realises the essential stupidity of
war and the infinite value of love.*

The Game of Life, detail

Battery, detail

'No one plays more passionately than an artist.
An artist who is travelling along a parallel-perpendicular path
To what? From what?
Nothing!
Glass: A touch of silica, fire, breath.
And then, what?

Mother nature provides, artists create – within a vast realm of possibility.  It seems as
though Peter Layton conjures his ingenious creations from all these strands.

Peter equals glass – This much is clear!'

65

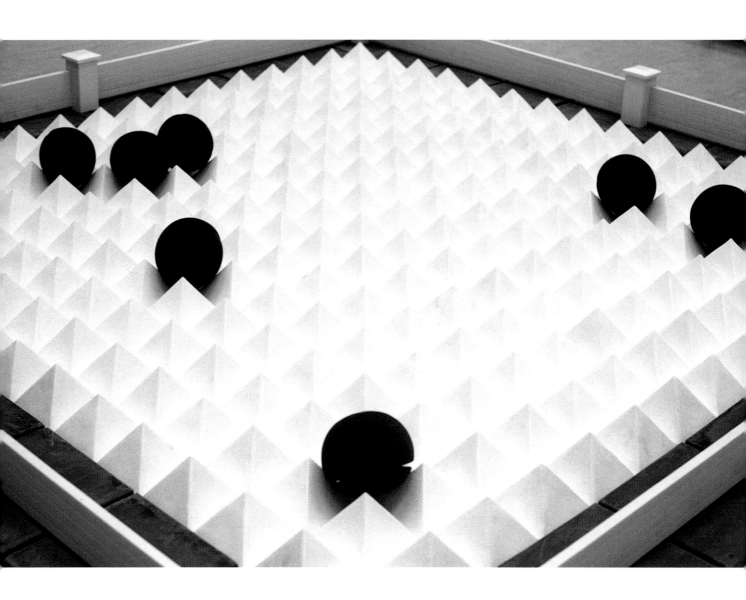

Bowling green, plaster, plastic and flocked
ceramics, 1972, 275cm x 275cm wide –
Commonwealth Art Gallery 1972

Since meeting Peter I have learned a lot about glass. It is among the most difficult
of mediums, and one has to be very skilful as well as loving, to extract art from the
furnace.

Peter Layton, with his Czech origins, carries the tradition in his genes.  He creates glass
that is in turn both subtle and bold, and gives his objects unique qualities that distinguish
his vision from that of other artists.

Encountering Peter's work for the first time, it struck me that it appeared more
decorative than sculptural, but I was intrigued and wanted to see more, to look again
and to look closer because the abstract landscapes trapped in his glass suggested
interesting hidden fantasies.

Alchemy, Half Full – Half Empty, 2006,
60cm high – glass, blood

I have worked with many artists over the years and I am impressed by Peter's openness and humanity. Nor am I surprised that serious collectors of contemporary art are interested in his work because his sculptures have powerful themes and messages expressed with originality, and are undoubtedly of museum quality.

I am delighted to contribute a few words in honour of the celebration of his dedication to glass and it is with great enthusiasm and pleasure that the Alexia Goethe Gallery represents his work.

**Alexia Goethe**
2006

Unforeseen Consequences, 2004, 100cm wide – neon, glass and mixed media
*While accepting that there are some things we seem unable to prevent, like natural disasters and even wars, I realise that despite frantic, often hedonistic effort, there is precious little we can ultimately control. Life goes its own sweet way (or otherwise) regardless.*

# the team – work and statements

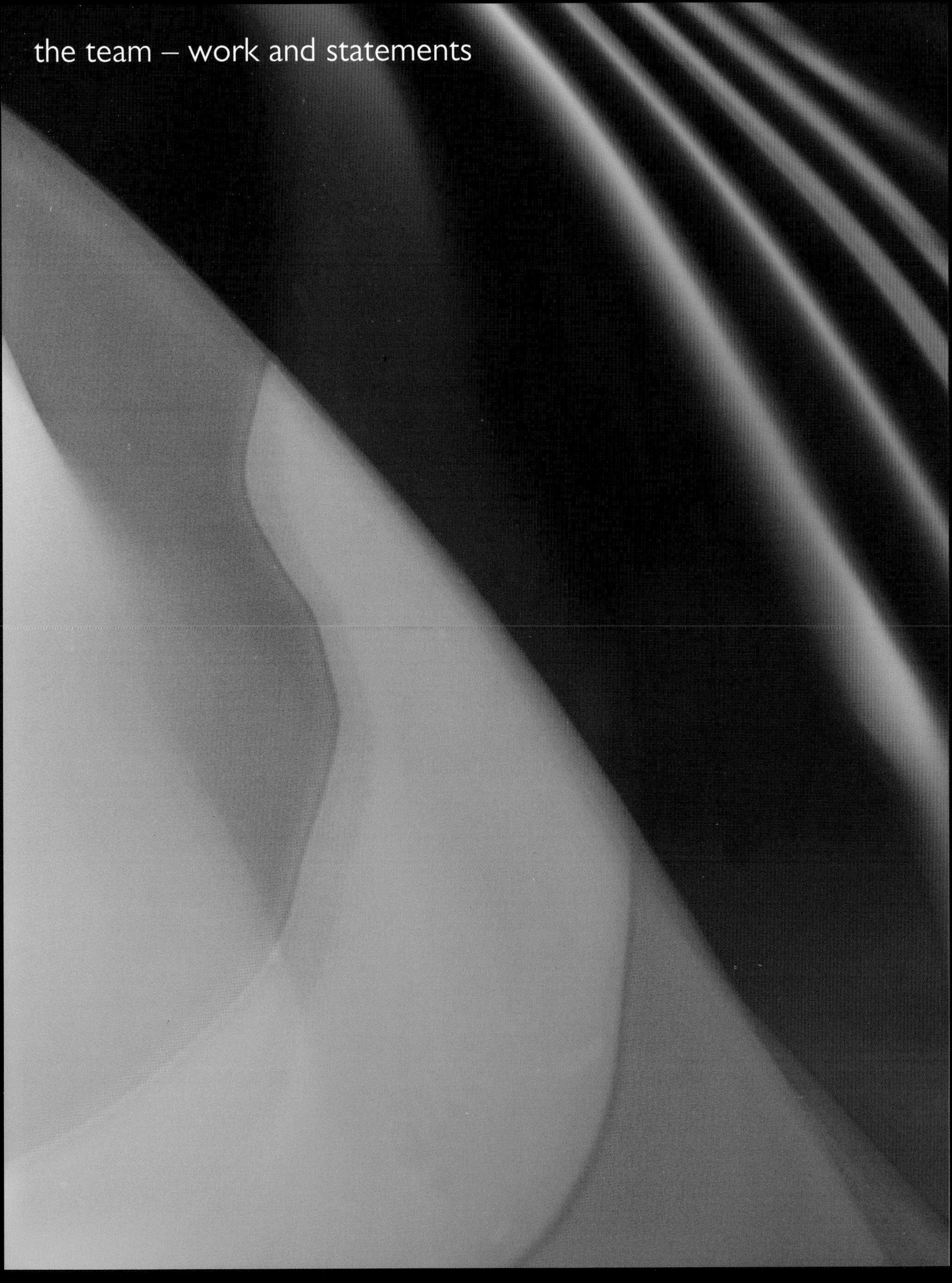

# sabrina cant

Aspects of nature sometimes become surreal in our everyday perceptions. Tranquil and mysterious, a scene in nature may draw you into its beauty. Evocative of childhood, the power of movement, colour and light propels you towards a magical world, taking you far beyond the boundaries of your garden; discovering one world through another, liberating the desire to explore and prompting all sorts of unforeseen adventures. The diversity of the landscape expresses many different things in a language that most of us can understand. We all experience nature. Most people relate to an experience of nostalgia, fear or happiness reflected through an encounter with landscape.

Painters have been the main source of inspiration for my glass. Artists have used the landscape as a canvas for their emotional responses to nature for hundreds of years. Kiln-formed glass, although a relatively new medium, has unprecedented potential in the area of abstract art. Glass with its natural qualities of inner space can be many things:

Colourscape. A day in the life of

Expanding cube, 18cm diameter

light, water and movement frozen in time, a controlled chaos. Rothko, his canvases steeped in colour, drawing us into a spiritual space, encouraged me to explore the qualities that glass has beyond painting, for example its three dimensional qualities. My fascination with Turner and Monet's development from realism into the abstract, using subdued washes of merging colour, helped me to clarify that using abstract colour is a comfortable language for me to express my own nostalgic feelings regarding space.

Initially, my progression into colour took its inspiration from landscape, but casting figuratively in clear glass with slight colour variations was a distraction from what I really wanted to achieve in my work. My studies at the Royal College of Art enabled me to realize that for me personally colour was the main factor in controlling form. The decision to stop including any literal elements in my work was the result of a need to work with colour in an abstract way. During my first year at the RCA I formulated and melted my own glass colours and documented all the results for future use. I could then control the density of the colour by varying the amounts of oxides. This gave me a good technical basis for understanding the nature of colour in glass, both during the developmental stages of my work, and for use in its future evolution.

My time at the RCA, creating and experimenting with pure colour, and my long-standing interest in painters who use abstract colour, resulted in a burning ambition to achieve in my glass a mysterious inner glow, with subtle changing effects expressing harmony, to conjure up an atmospheric illusion using minimal abstract forms. I use coloured glass to create an aura in space and time. A space that we do not just look at but become a part of, a space that we see idiosyncratically.

# bruce marks

I have a light-hearted relationship with glass; I would like the experience to stay the way it was when I first fell under its spell – magical, full of fun and excitement.  During my short acquaintance with glass I've learnt to let it guide me.  When I am open to all the hints and clues (secrets) that molten glass reveals - magic happens.  Sometimes I allow the glass to be shaped merely by heat and gravity, allowing it the freedom to create its own form.  My plan sometimes agrees with that of the glass, and at other times not at all, but each piece is a part of me, reflecting my thoughts, moods and feelings.

Nexus, 22 cm tall

Buoyant, detail

Flower petals, leaf veins and little details of nature really fascinate me. Nature seems to be able to create so effortlessly. Trying to capture these inspiring designs in glass will take me a lifetime and more. This is what drives me. It can be either very rewarding or extremely frustrating but never boring! Memories of my travels, the people I meet and their stories also inspire me.

Being part of the team is a huge advantage – to be able to discuss ideas, give and get advice and comments plays a key part in the development of my work, and I am really grateful for that. The learning never stops. Creativity breeds creativity.

I love glass. I dream glass. I talk glass. I think, breathe, make glass. I live glass. My life is full. Full of glass. The glass is half full – not half empty!

Buoyant, 12cm diameter

# simon moss

Peter and I have been collaborating since 1991 on a variety of works, many of which have been commissioned. We specialise in combining glass with metal to create pieces to enhance both public and private spaces. Exploring monumental scale, the dynamic interplay of light, colour and movement within the pieces and resolving the complex physical, spatial and aesthetic issues involved in the continuous search for daring and innovative solutions is a recurring and stimulating challenge.

Monolithic or magical, I strive to achieve a degree of visual ambiguity or mystery with kaleidoscopic images appearing and dissolving as the viewer moves and interacts with the piece.

Most recently, we have begun a series of cast pieces; a departure, which I feel, holds great potential for us to explore new skills, exciting ideas and endless possibilities.

Libra II, SM/PL, 2006, 30cm high
kiln cast glass, two intersecting glass
elements, polished and satin finishes

74

Opening, SM/PL © 2000,
Entrance sculpture – The Office Park,
Leatherhead, 6 metres high, 15mm
toughened float glass, stainless steel,
integral lighting

Opening SM/PL © 2000, Entrance sculpture detail –
The Office Park, Leatherhead, 6 metres high,
15mm toughened float glass, stainless steel, integral lighting

Synergy SM/PL © 2006, Mayer Brown
Rowe & Maw, suspended entrance foyer
stairwell sculpture, 4 metres high,
fabricated mirror polished stainless steel,
8 projecting bent and colour laminated
float glass sections

# yoshiko okada

Presently my work continues to explore the synergy and paradoxes of my Japanese background and experiences, and my current English and European situation. My key interest is in exploring memory, identity, time and the human condition and although seemingly complex this often leads me to simple forms of expression or symbolism.

Two face, 30cm high

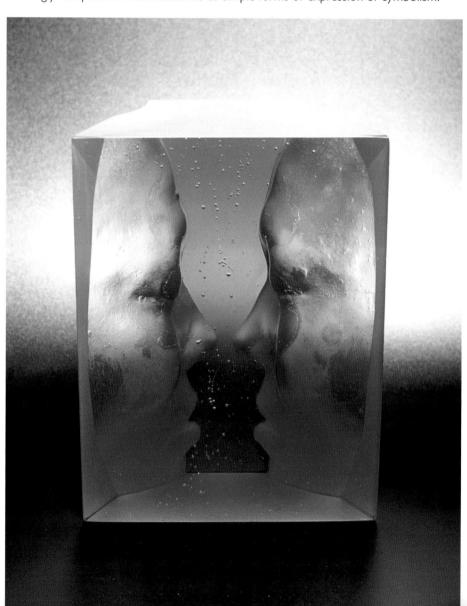

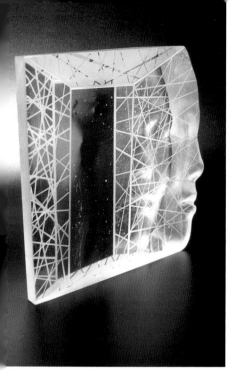

The missing piece I, 30 cm high

The techniques I use presently are kiln cast glass in life size sections of face and head with additions and subtractions, which are then detailed with photographically fused self portraiture and sandblasted imagery.

I am also trying to create a 'stage like' theatrical presentation in which a dialogue unfolds within the pieces of glass.

Dream I, 30 cm high

# layne rowe

My journey in glass began fourteen years ago during my degree course in three-dimensional design at the University of Central Lancashire.

For the next seven years I worked alongside a number of glassmakers at London Glassblowing Workshop before moving to Brazil for two years. Whilst there I set up a glass studio as well as experiencing a different working environment amongst industrial glassmakers. Since then I have run my own studio in Hertfordshire, a difficult and at times lonely existence, and I recently rejoined London Glassblowing to work together with artists and friends that I missed.

Scarify, detail

Scarify, 24cm high

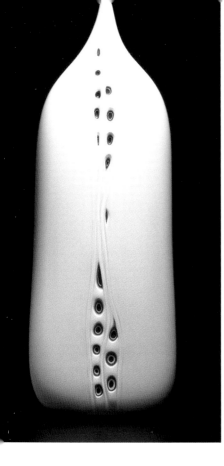

Scarify, 55cm high

My introduction to hot glass gave me an insatiable desire to explore this medium to its limits. I find glass, with its colour, physicality and myriad possibilities, an inspiration in itself. Further inspiration is drawn from natural forms and colours, the sea, the cosmos and the deep sense of mystery that one feels when overlooking water and vast open spaces.

Many of my pieces are a synthesis of techniques intended for a harmonious resolution, as if perfected by nature. The latest series, named Scarify, is created from a complex multi-coloured blank, overlaid transparent and opaque layers of colour with clear glass, which are then carved to create a textured surface exposing strata and veins similar to those of rocks eroded away by running water. This introduces areas of light and shade drawing the eye to the range of fine detail which together with new and experimental variations in shape and form, will benefit from repeated viewing.

Scarify, 22cm high

# anthony scala

I have worked with Peter Layton for a number of years and in that time I have come to know some exceptional glassmakers and witnessed the creation of some of the most vibrant and complex colour work to be seen in contemporary studio glass. My inspiration, however, is drawn elsewhere, from the complex optical effects that can be induced in glass, and how these can be manipulated to alter the perceived image of an object.

Untitled, 15cm diameter

Untitled, detail

For many people the world seems to exist almost entirely from information received through our eyes, but the notion that visual perception is the most susceptible of all our senses has been exploited by magicians and illusionists for hundreds of years. An accomplished illusionist exploits this implacable belief in what we see to perform seemingly impossible feats. Once the eyes have been deceived, our unwavering belief in physical laws, can for an instant, be called into question. This glimmer of uncertainty may be disturbing, yet it can also be utterly captivating.

Perception of an object is predicated upon light, angle and distance. Is what we see actually there or could it be a trick of the light? Take a step in any direction and the perceived image changes. After all, a holographic representation can appear entirely real until one reaches for it. These shifting perceptual qualities have come to dominate my work, providing an infinite source of inspiration.

Light exists without mass or substance, yet given the appropriate conditions, can generate images of absolute solidity. My work attempts to capture and utilise refractive elements to maximum effect, by engineering unorthodox optical structures which yield images that appear to distort the laws of three-dimensional space.

Untitled, detail

It is impossible to predict how the numerous optical relationships within each structure will interact with each other. I am constantly surprised by new and unexpected images created by internal reflections. These quirks of perception extend beyond pure geometry. The deeper you look into an object, the more there is to see.

81

# lucy swift

'There's an art which dominates by captivating, another which dominates by subjugating…a dynamogenic focus…rather than merely a transforming focus.' *

I have the opportunity to exercise my experimental thirst for glass through both teaching and private commissions. It is here that I exploit the notion of beauty, that which is general, discernable and pleasing. These activities are a valuable opportunity on which to build my technical knowledge as a glassmaker.

However, I have a different intention when focusing on my own work, based on a natural compulsion, which confronts and challenges the direct flow of my own sensibilities, which is particular and personal, derived from the spontaneous and organic. It may be felt by the spectator or not, but can never be overtly visible, because it is not ornamental.

Bricks group, 30 cm high

* Fernando Pessoa, notes for 'Estetica non-artistotelica' (1905-1920).

Grey Brick, 30 cm high

Aesthetically, I like to see my work as small structural fragments. To make them more than simply functional, vested with a finer register of value. For example, pieces are over long periods of time, added to, adapted, reconstructed, grouped, multiplied; each variable and connected within the context of their environment.

White Brick, 30 cm high

83

# louis thompson

Silent Fervour, 33cm long

Glass found me. I originally wanted to study Product Design but I was spellbound by the skill and energy of working with hot glass. After completing my BA Honours Degree in Glass at North Staffordshire Polytechnic I worked in a number of hot glass studios, collaborating with many different artists.

I returned to Staffordshire for nine years, following my appointment as Course Leader of the Department of Glass at Staffordshire University. I am now based at the London Glassblowing Workshop, where I continue to develop my work for exhibition both nationally and internationally.

My present body of work represents a time of transition. Exploring intimate spaces, suggestive of touch, soft and delicate impressions, sensual interaction and ambiguity. Hot glass enables me to investigate these themes and express my ideas through the

Songbirds, 34cm long

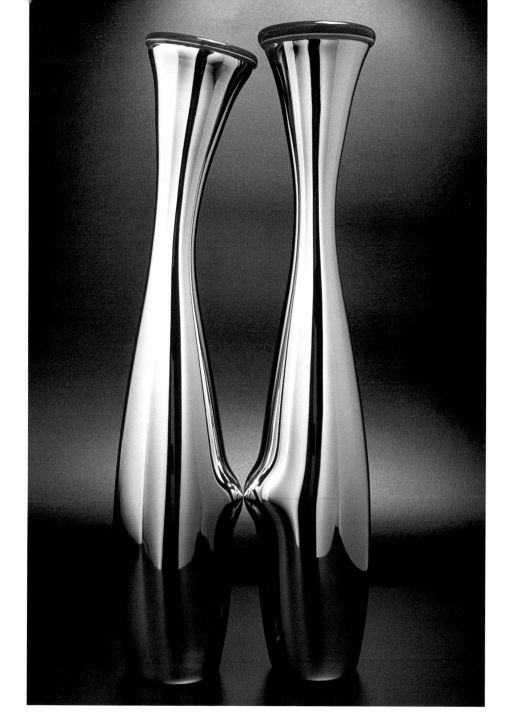

Those that never speak, 52cm high

visual language of form, colour and surface. My own curiosity is translated into objects for the viewer to explore. Refraction of light, opaque surfaces and a mirrored presence are aesthetic choices that dictate the distinctions between works, and express the beauty and dialogue of surface and form through the use of positive and negative space.

My passion for glass is informed by a deep fascination with the process of making, using hot glass as the medium of expression and enquiry. It is this process together with my particular vision that breathes life into the work in a way that is enigmatic and invites further scrutiny.

# marie worre hastrup holm

In my current work I want to make people question their expectations, and more particularly their preconceived ideas about 'glass'.

What purpose should a glass object serve? When is an item no longer functional? In relation to our viewpoint and movements within a space, where do we presume a certain item should be and why?

On encountering an object, what determines our understanding of it? What defines how we perceive it even before we consciously consider it?

Addressing these issues, I blow easily recognisable, regular shapes to which I add various elements in glass and other materials, in ways that might take us by surprise and make us smile.

Bloomin' Bowl, 24 cm diameter

Bloomin' Bowl, 24 cm diameter

Bloomin' Bowl, 24 cm diameter

# afterword

Paradiso green stone form, 2004,
20cm high

'THE TIMES THEY ARE A CHANGING!' – Bob Dylan

The past is full of rich and heart warming memories of the contemporary glass world in its infancy, developing and growing into an extended international family, replete with extraordinary and generous characters, some of whom are sadly no longer with us. The charming, gracious and unquestionably great Stanislav Libensky, the hard-drinking avant-garde Joachim Torres-Esteban, the dedicated priest turned obsessive glass-lover Louis Meriaux, and other visionaries like Sybren Valkema, Roberto Neiderer and many more who contributed with such passion in those early days of European studio glass. The pioneering generation is moving right along, the fathers of the Studio Glass Movement have become the great grandfathers – people like Erwin Eisch, Sam Herman, Finn Lynggaard, Willem Heesen and Claude Morin. Not to mention those on the other side of the pond – Harvey Littleton, who set the ball rolling, his protégées Marvin Lipofsky, Fritz Dreisbach and Dale Chihuly, along with early converts Joel Philip Myers, Henry Halem and countless others.

Those really were exciting times – the incredible working symposia in Novy Bor, Frauenau, Sars Poteries, London, Lucerne, Rouen, Japan, Russia and elsewhere, where the intense atmosphere of learning and sharing, if occasionally competitive, transcended all barriers of language, culture and politics. In Britain it has been a long haul to create a viable audience for contemporary glass – a continuous struggle against the tide of blinkered appreciation of little other than traditional cut lead crystal despite which the domestic industry has reached terminal decline. And yet today, we see glass art beginning to have some impact in reversing this trend. Collectors are starting to take notice and respond, albeit modestly, creating a demand and a corresponding rise in prices for the more exciting and accomplished works. Although public interest still lags behind other countries – with many of our best artists, including Colin Reid, David Reekie, Anna Dickinson and Tessa Clegg, being forced to show and sell almost exclusively abroad, exhibitions like Chihuly at Kew and the British Glass Biennale in Stourbridge are helping to raise the profile of the medium.

For the future, I look forward to a situation in which glass art transcends the inter-
minable art versus craft debate; no longer a Cinderella medium, but sitting comfortably
alongside other art forms.  The discussion will then turn on whether the art is good or
bad rather than the hierarchy of media.  Naturally there will be those who will continue
to love the material for its inherently seductive qualities but others will value its unique
potential to contain and express conceptual and sculptural possibilities.

I remember some years ago a forum discussion where it was suggested that glass artists
might one day find their work presented in the Tate (our foremost museum of Modern
Art), to which came the tart rejoinder that we should be content to aim for the Victoria
& Albert Museum, (decorative and applied arts).  Absolutely nothing wrong with that –
a fine museum, but it is nevertheless encouraging to note that attitudes really do seem to
be changing.  Dan Klein has written in the Coburg Prize catalogue 'glass has become
accepted worldwide as a valid vehicle for self expression in the 21st century.  This in itself
has brought about major change.  Glass is now used unquestioningly for sculpture, two
dimensional art, installation art, conceptual art, architecture and most recently video'.

89

While this is unlikely to be the last word on the matter, I suggest that it shows a
considerable shift in the right direction – towards glass as a fully-fledged and
'emancipated' medium.

**Peter Layton**
2006

# artists'
## CURRICULUM VITAE

## Peter Layton

Education
1962-65    Central School of Art and Design, London
1960-62    Bradford College of Art, England

Work Experience
2006       The Workshop is 30 years old, the oldest glassblowing
           studio in the UK and one of the first in Europe
1996       Publication of *Glass Art* by Peter Layton
1976       Established the London Glassblowing Workshop

           Lecturer: Various Art Colleges in USA and UK,
           including Camberwell, Croydon and Hornsey
           Colleges of Art

Awards
2003       Honorary Doctorate of Letters from the
           University of Bradford
2002       Honorary life member of the Contemporary
           Glass Society

Selected Exhibitions
2006       Peter Layton and Friends, Ruskin Glass
           Centre/World of Glass
           Peter Layton and Peter Bremers, Raglan Gallery,
           Australia
           Peter Layton and Peter Bremers, Freed Gallery, USA
           Peter Layton solo show, Alpha House Gallery, UK
           Group Show, Alexia Goethe Gallery, London
           Peter Layton and Friends, Candover Gallery
           Tangents, Glass Art Gallery, London
           Peter Layton and Friends, Church Gallery, Chagford
2005       SOFA Chicago, USA
           Galerie Klute, Germany
           Leon Salet Gallery, Maastricht, Holland
           Glass Trail/Chihuly Gardens of Glass, Kew, UK
           London Art Fair, London, UK
           21st Century British Glass, London, UK
2004       British Glass Biennale, Stourbridge, West Midlands
           Fragile Cargo, Anglo Hungarian travelling exhibition
           Art in Essence, London, UK

Metamorphosis, London, UK
Isle Gallery, Isle of Man, UK

2003       Feel the Heat', Grace Barrand Design Centre
           The art of glass and embroidery', The De Morgan
           Centre
           Glass Inspirations, Switzerland
           Vessels, Romney Marsh Craft Gallery
           Six of the Best, Candover Gallery
           Heavenly Scent, The Glass Art Gallery, London
           Haiku and Glass, The Glass Art Gallery, London

Public Collections
           Fitzwilliam Museum, Cambridge
           Liverpool Museum
           Castle Museum, Nottingham
           National Museum, Prague, Czech Republic
           Victoria and Albert Museum, London, England
           Bradfield House Glass Museum
           Gus Knrvstainy Glass Museum, Russia
           International Glass Museum, Ebeltoft, Denmark
           Glasgow Museum
           Norwich Museum and Art Gallery
           Royal Scottish Museum, Edinburgh
           Leicester Museum, UK

Selected Commissions
2006       Short listed for 'Gateway to Yorkshire' commission
           Sentinels – MV. Talisman cast glass and metal niche
           pieces
           Synergy – Suspended sculpture (Mayer Brown Rowe
           and Maw)
           Elios – maquette (Cass Sculpture Foundation)
2004       Axios – Suspended, rotating atrium sculpture, Apex
           Plaza
2002       Cruise line sculptures, Celebrity Cruises (RCL)
           Princess Lines
           Tree of Life, A Picture of Health Limited (Sculpture)
2001       Opening – Leatherhead Office Park sculpture

## Sarina Cant

### Education

|  | MA Ceramics and Glass, The Royal college of Art |
|  | BA (hons) Ceramics and Glass, Buckinghamshire University |
| 1994-1996 | Braintree College, GNVQ AVD Art and Design |

### Work Experience

| 2006 | Peter Layton and Associates |
|  | 'Glassforms' Assistant to Max Jacquard |
| 2004-Present | Visiting Lecturer, BCUC, Ceramic and Glass, High Wycombe Campus |

### Awards

| 2004 | The New Designers Bombay Sapphire Award for Contemporary Glass |
|  | The Thames and Hudson RCA Society Art Book Prize |
| 2001 | The Glass Sellers Student Award 2001 |

### Selected Exhibitions

| 2006 | Peter Layton and Friends, Church Gallery/Ruskin Glass Centre/World of Glass |
| 2005 | 'One Year On' The Business Design Centre |
| 2004 | 'The Bombay Sapphire Prize' 2004 touring exhibition |
| 2003 | 'Transfuse' The Great Eastern Hotel, London |
| 2002 | 'Henley Festival 2002' Buckinghamshire |
|  | 'Art 2002' Business Design Centre, London |

## Bruce Marks

### Work Experience

| 2003 – present | London Glassblowing Workshop |
| 1998 | Freelance lampworker, jewellery designer |

### Exhibitions

| 2006 | Peter Layton and Friends, Ruskin Glass Centre/World of Glass |
|  | Spring Show, Obsidian Gallery, Stoke Mandeville, UK |
|  | Peter Layton and Friends, Church Gallery, Devon, UK |
|  | Group show, Chapel Gallery, Bedford, UK |
|  | Tangents, Glass Art Gallery, London, UK |
| 2005 | Glass Trail, Chihuly at Kew, UK |

### Professional Memberships

| 2006 | Glass Art Society, USA |
| 2005-2006 | Contemporary Glass Society, UK |

### Commissions

| 2006 | Memories, private commission, lampworked |

## Simon Moss

Education
1986-89    Ravensbourne College of Design and Communication
           BA Hons Degree Product Design
1985-86    Somerset College of Arts and Technology

Selected Exhibitions
2006    Peter Layton and Friends, Ruskin Glass Centre/
        World of Glass
        Church Gallery, Chagford Devon
2005    Shaping Light 2, Workplace Art
        Glass in the Garden, Waterperry Gardens
2001    Yorkshire Craft Centre
1999    Glassworks, Bluecoat Display Centre
        Interior Designers and Decorators Association
1998    Glass UK, The National Glass Centre

Selected Commissions
2006    Short listed for 'Gateway to Yorkshire' commission
        Sentinels – MV.Talisman cast glass and metal niche
        pieces
        Synergy – Suspended sculpture (Mayer Brown Rowe
        and Maw)
        Elios – maquette (Cass Sculpture Foundation)
2004    Axios – Suspended, rotating atrium sculpture Apex
        Plaza (Prudential)
2002    Pool deck sculptures, Celebrity Cruises (RCL)
2001    'Opening' Leatherhead Office Park sculpture
        (Frogmore Development)

## Yoshiko Okada

Education
2001–2004    University College for the Creative Arts
             BA (Hons) Three Dimensional Design Degree in Glass
2000–2001    South Thames College, London
             Decorative Glass: City and Guilds Decorative
             Glass Part 2
1997         Morley College, London
             Glass Engraving trained under Peter Dreiser
             Ceramic trained under Gill Crowley

Work Experience
2001–2006    London glassblowing, Peter Layton's studio, London
2004         Residency at Northlands Creative Glass, Scotland
2000–2002    London Stained Glass Co., London

Awards
2004    First Prize for Worshipful Company of Glass Sellers
        Student Award, London
        Short listed for Bombay Sapphire Martini Glass Prize
        competition, London
2001    Short listed for Stevens Glass Competition, London

Exhibitions
2006    Peter Layton and Friends, Church Gallery/Ruskin
        Glass Centre/World of Glass
        British Glass Biennale, West Midlands
        New Glass 06, Cowdy gallery, Gloucestershire
        Coburg Glass Prize 2006, Germany
        Contemporary Glass from Japan, Plateaux Gallery,
        London
        Transparent Rock, Oxfordshire
2004    Mixed show, Northlands Creative Glass, Scotland

## Layne Rowe

Education

BA Hons, 3D Design, University of Central Lancashire UK
BTEC National Diploma –Art and Design, Shephalbry College,
Stevenage, UK

Work Experience

| | |
|---|---|
| 2005 | Rejoined London Glassblowing |
| 2003 | Established Layne Rowe Glass Studio |
| 2002-03 | Freelance work for a number of established glass studios |

Lived and worked in Brazil setting up studio for hot glass 'Crystalaria
Guanabara' glass factory

| | |
|---|---|
| 1999-2001 | London Glassblowing |

Exhibitions

| | |
|---|---|
| 2006 | Peter Layton and Friends, Ruskin Glass Centre/ World of Glass |
| | Tangents, Glass Art Gallery, London |
| 2005 | Leon Salet – Maastricht, Holland |
| | Glass Trail – Chihuly at Kew |
| 2004 | British Glass Biennale Stourbridge |
| 2003 | Apple Gallery, Goddalming, Surrey |
| 2002 | CASA COR 2002, Rio de Janeiro, Brazil |

Commissions

| | |
|---|---|
| 2003 | World Idol Trophy, Thames Television |
| | CASA COR Claraboia (Sculpture for skylight) |
| | Nokia to Orange Award (Spiral elements) |

## Anthony Scala

Education

| | |
|---|---|
| 2001 – to date | Freelance Artist, Designer and Maker Specialising in cold glass – working and assemblage techniques |

Work Experience

| | |
|---|---|
| 1999 – 2003 | London Glassblowing Workshop Glass Design Apprenticeship |
| 1996 – 1998 | Kent Institute of Art &Design HND Modelmaking |

Awards

| | |
|---|---|
| 2005 | Main Award, Glass Sellers Prize |

Exhibitions

| | |
|---|---|
| 2006 | Peter Layton and Friends, Ruskin Glass Centre/ World of Glass |
| | Tangents, Glass Art Gallery, London |
| 2005 | Glass Sellers Exhibition, Principal Prize Winner London |
| | London Glassblowing Exhibition, Leon Salet Gallery, Maastricht |
| 2004 | British Glass Biennale, The Ruskin Glass Centre Stourbridge |
| | Metamorphosis, Zest Gallery, London |

Commissions

| | |
|---|---|
| 2006 | Glass Sellers Technical Achievement Trophy |

## Lucy Swift

Education

Education

| | |
|---|---|
| 2004 | Birkbeck University – Diploma: Psychoanalysis and Art |
| 2000 – 2002 | University of East London – MA: Art In Architecture |
| 1993 – 1997 | Edinburgh College of Art – BA (Hons): Design and Applied Arts – Glass and Architectural Glass |

Work Experience

| | |
|---|---|
| 2006 – Present | Established Studio, London Undertaking Commissions and Private Tuition |
| 2003 – Present | The Art Academy, London, Resident Artist/ Glass Tutor |
| 2002 – Present | London Glassblowing Workshop/Peter Layton & Associates/the Glass Art Gallery, Gallery Co-ordinator |
| 2001 – 2005 | Core Team Member of Art & Architecture |
| 2000 – Present | Amy Cushing – Mosquito Glass Design/ Isabelle Starling, Studio Technician |

Selected Exhibitions

| | |
|---|---|
| 2006, 2005 2004, 2003 | Resident Artists' Show, The Art Academy, Union Street, London |
| 2003 | Glas, Glas Gallery, Broadway Market, London |
| 2002 | New London Glass, Espai Vidre Glass Gallery, Barcelona, Spain |
| | Tracks-New London Glass, Gloucester Road Underground Station, London |
| 2001 | East London Design Show, Shoreditch Town Hall, London |

Selected Commissions

| | |
|---|---|
| 2006 | Levis' Strauss, Private Commission. Three fused, cast and carved crystal glass sculptures |
| | Alixandra Beadale, Artist. Five cast crystal glass figures /portraits |
| | Total Theatre, Edinburgh Festival Awards. Twelve cast glass, cut and polished sculptures |

## Louis Thompson

Education

| | |
|---|---|
| 1985–1988 | B.A. Honours Degree in Glass Design North Staffordshire Polytechnic |

Work Experience

| | |
|---|---|
| 2006 | Visiting Lecturer, University of Sunderland Senior Lecturer and Joint Course Leader for BA Honours Design in Glass, Staffordshire University, UK |
| 2001 | Teaching Assistant, Pilchuck Glass School, Seattle, USA |
| 2001-2006 | Freelance Glassmaker London Glassblowing Workshop, London |
| 1990–1996 | Production Glassmaker, E & M Glass, Wales |

Awards

| | |
|---|---|
| 2001 | Saxe Award Nominee, Pilchuck Glass School, Seattle, USA |
| 2000 | Artist Residency, Leerdam Glass Centre, Netherlands |
| 1999 | Scholarship Student, Pilchuck Glass School, Seattle, USA |

Selected Exhibitions

| | |
|---|---|
| 2006 | Peter Layton and Friends, Church Gallery, Ruskin Glass Centre /World of Glass |
| 2006 | International Glass Biennale, Stourbridge (August) |
| 2005 | Four Person Show, Leon Salet Gallery, Maastricht, Netherlands |
| 2004 | British Glass, Glass Inspiration, Burgdorf, Switzerland |
| 2003 | Korean World Craft Biennale, Cheongju, Korea |
| 2000 | Vessels, International Glass Exhibition, Kanazawa, Japan |

Commissions

| | |
|---|---|
| 2006 | Interior Installations with artist Clem Crosby Artist, Young Vic Theatre, London |
| 2005 | Glass Sculpture for artist Clem Crosby, London |
| 1999 | Bluecoat Display Centre Refurbishment, Liverpool |

picture credits

Mireslav Vojtechovsky: p33

Lumir Rott: Pyramid p31, Hurdle p34

George Erml: Endless column p36

David Cripps: Morar vase p26

**Marie Worre Hastrup Holm**

Education
| | |
|---|---|
| 2000–2003 | BA(Hons) Visual Arts, Painting, Camberwell College of Arts, London |
| 1999–2000 | HNC Fine Art, Kensington & Chelsea College, London |
| 1992–1993 | The International Glass Centre, Brierley Hill, Dudley College of Technology |

Work Experience
| | |
|---|---|
| 1996–2006 | Production of own designs comprising abstract and functional work and freelance glassblower for Peter Layton, London Glassblowing |
| 2005 | Designer, co-organiser and curator of the exhibition Wearing Glass at the.gallery@oxo, London and the National Glass Centre, Sunderland |
| 2004 | Glassblower at Gler i Bergvik ehf glass studio, Sigrun O. Einarsdottir, Iceland |

Awards
| | |
|---|---|
| 2006 | Artist in Residence at Corning Museum of Glass, USA |
| 1998 | Scholarship in drawing and painting at Oliver Bevan's studio school |

Selected Exhibitions
| | |
|---|---|
| 2006 | Peter Layton and Friends, Church Gallery/Ruskin Glass Centre/World of Glass |
| | Tangents, Glass Art Gallery, London |
| 2005 | Plumbline Gallery, St. Ives, Cornwall |
| 2004 | British Glass, Glass Inspiration, Switzerland |
| | British Glass Biennale 2004, The Ruskin Glass Centre, Stourbridge |
| | What's New… Danish Glass 2004, Ebeltoft Glass Museum, Denmark |
| 2003 | Selected to exhibit on the 'One Year On' Crafts Council at New Designers in the Business Design Centre in Islington |

Commissions
| | |
|---|---|
| 1999 | DW Productions UK Ltd., oil lamps for the film *Gladiator* |

Martin Andrews

Joanne Cailes

**Sabrina Cant**

Anna Chrysopoulo

Norman Stuart Clarke

Anna Dickinson

Julia Donnelly

Carin Von Drehle

Cindy Elliott

Candice-Elena Evans

David Flower

Claire Gutteridge

Anna Hazelden

Geoff Innell

Steve Jackson

Max Lamb

Siddy Langley

Karen Lawrence

**Bruce Marks**

**Sylvie Marks**

Gayle Matthias

Mike McGregor

Carrie McKnight

**Simon Moss**

Aileen Newton

**Yoshiko Okada**

Thomas Petit

Richard Price

**Layne Rowe**

**Anthony Scala**

Lesley Scholes

Killian Schurmann

Tamlyn Smithers

Patrick Stern

**Lucy Swift**

**Louis Thompson**

Tina Tylen

Charles Victor Ramsay

Tim Waldegrave

Jessica Watson

David Weeks

**Marie Worre Hastrup Holm**

Alisa Yemini

* Names which appear in bold denote current members